Birds in Flight

edited by
Roger Caras

foreword by
Roger Tory Peterson

Westover
Publishing Company
An affiliate of Media General, Richmond, Va.

Prepared in cooperation with
Photo Researchers, Inc.,
New York, New York.

Book design by Sylvan Jacobson

Copyright® 1971 by Westover Publishing Co.
First Edition Library of Congress Catalogue
Card Number 75-161078. All rights reserved.
This book, or parts thereof, must not be
reproduced in any form without written
permission from Westover Publishing Co.

SBN 0-87858-012-3

INTRODUCTION

Few wonders touch man as deeply as the power of natural flight. Our mechanical approximations are coarse by comparison and we are well aware of it. Natural flight joins creature and sky. There is no intrusion. No bird is foreign to the wind.
When we lift ourselves away from the earth it is our superb intelligence that is putting us there but there is not intelligence enough to find a way in for us. We can never really join the sky. We understand many of its forces and we can call upon them with steel and glass and tons of wire and rivets. But our sense of alienation is never alleviated.

We have been watching the birds since before we were men. They flicked their shadows across our path with elegant disdain when we were yet without language. We listened to the wind in their feathers from our perches in trees. And when we progressed enough we marveled at their beauty of color and form. Man has always wondered about the bird. It would be a tragedy beyond measure, far beyond comprehension, if the course we have set should leave us nothing to wonder about but ourselves. Indeed, then, man might well wonder about man. In the meantime prayer would not seem out of order. We might well pray that our intelligence not have as its end product the elimination of beauty from earth, and bird from sky.

          Roger Caras

FOREWORD

The main attraction of birds is that they fly; they symbolize the ultimate in freedom and mobility. But they are not the only living things that have freed themselves from earthy fetters. Most insects, from tiny midges to psychedelic butterflies, can fly. Bats, which are mammals, have also evolved true flight, far more sophisticated than the rudimentary gliding flight of flying squirrels. Even man, within the present century, has taken to the air in those contrived extensions of himself, the airplane and the glider.

However, birds have a unique advantage; they are the only creatures with feathers. Because of the feather, a marvel of natural engineering, they are the most efficient aeronauts of all. They may not fly as high as aircraft, nor as fast, but they are far more maneuverable and consume less fuel.

Not all birds fly. Ostriches don't, nor do about 45 other species representing a dozen families, but all have descended, we believe, from ancestors that could fly. Penguins, for example; they still retain a keeled breastbone and powerful pectoral muscles. Indeed, we can argue that they still *do* fly, but in a heavier medium, for they use their wings, not their feet, for submarine propulsion.

Bird flight takes many forms, from the dynamic soaring of the albatross over the sea, the static soaring of a hawk on a thermal, the burst of speed of a flushed grouse, and the long-distance endurance of a plover or a tern to the phrenetic, buzzy hovering of a hummingbird.

Functionally, flight has enabled these attractive, incisive creatures to populate nearly every square mile of the earth's surface except the uttermost polar regions.

                                        Roger Tory Peterson
                                        Old Lyme, Connecticut

Long-tailed Jaeger

There are highways in the sky, with an endless number of lanes and a compass set to every degree in the full circle. There are layers as well as lanes, levels and slopes that ascend and descend invisibly but true. And to all of this magnificent architecture of the sky the bird attunes its whole body.

Every feather is at work, every muscle, every sinus that gives it lightness. The whole creature welds itself to the molecules of air, and bird and sky become one. By contrast we belch and blunder through in our machines and pray the air will not drop away and leave us stranded a mile from earth.

Mallard Ducks
"Although all ducks fly in the same style, each species has its little idiosyncrasies, whereby an experienced sportsman is able to distinguish it when on the wing. It is possible to identify each species merely by the sound made by the wings of a flock in motion. Few naturalists are able to do this, but very little experience enables a sportsman to see at the first glance whether or not a flying bird is a duck. Except when about to alight, a duck never sails on outstretched pinions, nor does it ever hover kestrel-like. It flies swiftly, with steady, rapid wingbeats, the neck being fully extended. Imagine a sodawater bottle on wings, and you have a fair mental impression of the form of a duck as it speeds through the air."
> From *Game Birds* by
> Douglas Dewar
> (Chapman & Hall, London, 1928)

Secretary Birds

Secretary birds have a strange effect on people. They are silly looking birds, and as they do their cakewalk through the African grass trying to turn up snakes and lizards, people somehow think of them as pint-sized ostriches. They aren't, not at all. And people shouldn't be surprised when they suddenly take wing or come soaring in over one's head. Although they spend a lot of time on the ground—if you eat snakes that's where you have to be—secretary birds are nothing more or less than very long-legged eagles. They nest high and fly well. It just so happens that they walk well, too.

Mallards on the Wing

"Look at that mallard as he floats on the lake; see his elevated head glittering with emerald green, his amber eyes glancing in the light! Even at this distance, he has marked you, and suspects that you bear no good-will towards him, for he sees that you have a gun, and he has many a time been frightened by its report, or that of some other. The wary bird draws his feet under his body, springs upon them, opens his wings, and with loud quacks bids you farewell."

> From *The Birds of America* by
> John James Audubon
> (1840-1844)

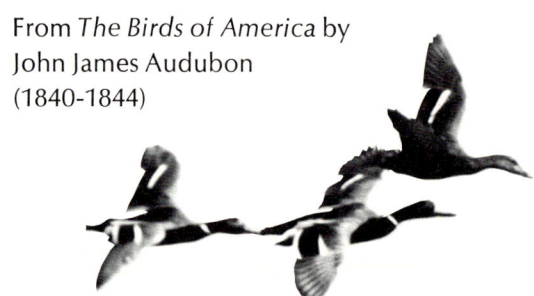

Vulture

It's strange what flight can do for an unlovely creature like a vulture. On the ground a vulture is most unattractive. Its habits are repulsive to most people largely due to their ignorance. A vulture is actually very clean, and very necessary in nature. But, dirty or clean, ungainly, bald, quarrelsome, all of this somehow doesn't matter when a vulture is on wing. Its soaring flight is the epitome of grace and ease. It relaxes on the wind and is at home in the skies. Strange, on the ground amid carnage a vulture seems a thing of the devil. Yet, on wing, it gets very close to Heaven. It would be funny if that is all that stood between the devil and an angel—the power of flight.

Barn Owl

An owl in flight comes as close to being a giant moth as anything. It is very un-bird-like. It does have to flap its wings, of course, but somehow it isn't a bird you are seeing but a gigantic soaring insect. There is a science-fiction quality to an owl. I remember seeing one circling in the light of a street lamp once. It was eerie. In some strange way the power of the bird came through. The bird was arrogant, I think. The owl was arrogant because it knew its own power and that there was death for a little animal whenever the feathered tiger took to the night sky.

Gulls

There can be no doubt that the flight of birds has a great deal to do with our intelligence. Our imagination drew us out of the fog of ape intellect and into the bright sun of human potential. Our imagination, in turn, had strings attached. The birds ran to the sun, to the moon, to the inaccessible mountain top, into the sea beyond the breakers, over the ridge, into the earth, through the pores of time, across the texture of wonder, and under the scrutiny of eyes becoming less misty as each million years slipped away noted only in the memory of the rocks. And there was one other thread, a golden one, and its end was held in the beak of every bird that lifted free of the land and floated on a cushion of awe.

Stork in Flight

Bald Eagle

"If one may believe the farmers' accounts, they add, that he will attack a deer sometimes: in this enterprise he makes use of this stratagem; he soaks his wings in water, and then covers them with sand and gravel, with which he flies against the deer's face, and blinds him for a time; the pain of this sets him running about like a distracted creature, and frequently he tumbles down a rock, or some steep place, and breaks his neck; thus he becomes a prey to the eagle."

    From *The Natural History of Norway* by
    The Right Reverend Erich Pontoppidan, Bishop of
    Bergen, Norway (1755)

Pigeon Hawk

"*King Henry:* But what a point, my lord, your falcon made, And what a pitch she flew above the rest! To see how God in all his creatures works! Yea, man and birds are fain of climbing high.
Suffolk: No marvel, an it like your majesty, My lord protector's hawks do tower so well; They know their master loves to be aloft, And bears his thoughts above his falcon's pitch.
Gloucester: My lord, 'tis but a base ignoble mind That mounts no higher than a bird can soar."

From *King Henry VI* by
William Shakespeare, Act 2.

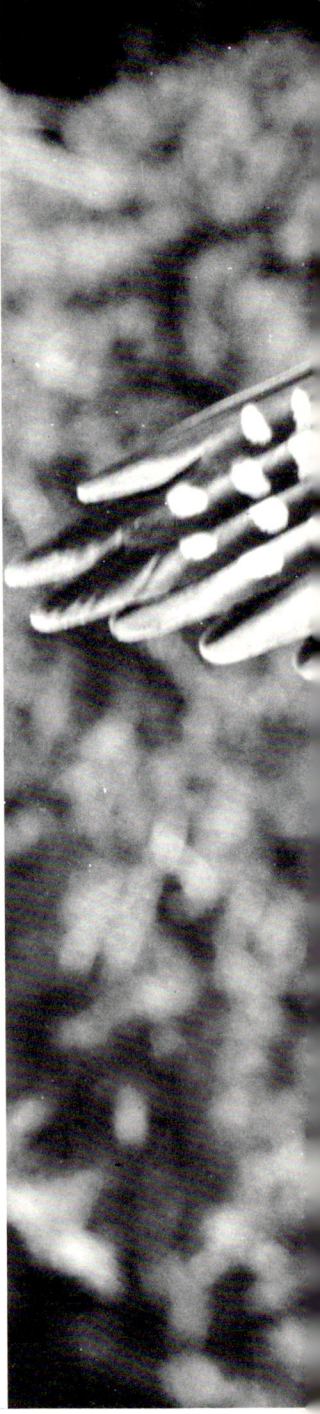

Tufted Titmouse

Galapagos Hawk

"Why are the bones of birds hollow? Because they are thereby rendered lighter, and do not interfere with the flight of the bird as they would do if they were solid. Greater strength is also obtained by the cylindrical form of the bone, and a larger surface afforded for the attachment of powerful muscles. Why do all birds lay eggs? Because, to bear their young in any other manner, would encumber the body, and materially interfere with their powers of flight."

From *Forest and Jungle or Thrilling Adventures in All Quarters of The Globe* by P. T. Barnum (1903)

Trumpeter Swan

"... when it is seen smoothly sailing along the water, commanding a thousand graceful attitudes, moving at pleasure without the smallest effort; 'when it proudly rows its state,' as Milton has it, 'with arched neck, between its white wings mantling,' there is not a more beautiful figure in all nature. In the exhibition of its form, there are no broken or harsh lines, no constrained or catching motions; but the roundest contours, and the easiest transitions; the eye wanders over every part with insatiable pleasure, and every part takes a new grace with a new motion."

From *A History of the Earth and Animated Nature* by Oliver Goldsmith

Canada Geese

Mute Swans

There is a special character a bird assumes when it is of the two worlds of sky and water. Somehow, waterfowl seem bigger and stronger, more powerful and mysterious. They are larger than the average bird, of course, and they migrate for the most part; therefore they are powerful. A robin migrates, too, but it isn't really very mysterious. Water birds, though, seem to come and go. There is a now-you-see-them, now-you-don't

quality to them. They are always either coming from someplace far away or heading there. They wedge into formation and set off to see the world. They fly high, and below we wonder about them, and worry sometimes, and listen for their voices. A world without them is unthinkable and, unfortunately, forseeable. We just might wipe them out yet!

Fairy Tern

Sea birds have an indestructible quality other birds don't have. Perhaps this is because of the eternal quality of the sea itself. We return to the sea, most of us, periodically, and the sea birds are always there just exactly as they were the last time, and the time before that. They perch in the same places, wheel through the sky on the same invisible avenues of wind, make the same sounds. It wouldn't be the coast, the sea, the boat, or the pier without them. Intellectually we understand that birds are not immortal, that on each visit we are seeing different birds. But, emotionally we don't understand this at all. The terns and the gulls are the same—they have always been there skillfully painting their wind pictures with their wing tips and they always will be. We change; they don't.

Black Skimmers

Peregrine Falcon

"A thousand feet high, he poised and drifted, looking down at the small green fields beneath him. His body shone tawny and golden with sunlight, speckled with brown like the scales of a trout. The undersides of his wings were silvery; the secondaries were shaded with a horseshoe pattern of blackish bars, curving inwards from the carpal joint to the axillaries. He rocked and drifted like a boat at anchor, then sailed slowly out onto the northern sky."

From *the Peregrine Falcon*
by J. A. Baker
(Harper & Row, New York, 1967)

Frigate Bird

We are all pirates at heart. There is not one of us who hasn't a little larceny in his soul. And which one of us wouldn't soar if God had thought there was merit in that idea? So, when we see one of those great widespread pirates soaring across the grain of sea winds we thrill, and we long, and, if we are honest, we curse that we must be men every day. Why not one day a bird! There's an idea, now, one day out of seven a pirate in the sky. What puny power a man **can** attain by comparison. Compare a 747 with a bird and blush!

Laughing Gulls

"The very idea of a bird is a symbol and a suggestion to the poet. A bird seems to be at the top of the scale, so vehement and intense in his life.... The beautiful vagabonds, endowed with every grace, masters of all climes, and knowing no bounds—how many human aspirations are realised in their free, holiday-lives—and how many suggestions to the poet in their flight and song!"

From *Birds and Poets* by John Burroughs (1887)

Osprey or Fish Hawk

"It is interesting to watch the fish hawk obtaining its food. Sailing along from 50 to 100 feet above the water, with its keen eye it can easily see any fish swimming close to the surface of the water, and as soon as it sees its quarry, stops its flight, remains suspended motionless in the air for a moment, closes its wings, and then darts downward like an arrow. It disappears under the water for a few seconds, and when it rises and again takes wing a shining, wriggling fish can be plainly seen in the grasp of its powerful talons."

    From *Life Histories of North American Birds*
    by *Charles Bendire* (1892)

Steller's Sea Eagle

American Egret
Animals gave us art. It was they who inspired us. The first graphics—petroglyphs and cave paintings—were of animals. And if you want to know where the dance came from watch an egret sometimes. We discovered the beauty inherent in our own bodies by watching birds dance away from earth. Too bad we can't hear the suites they hear. Especially when you consider who their composer is.

Hummingbird

The wonderful thing about hummingbirds is that for them flight seems impossible. Their legs aren't strong so there is no jumping involved when they take off vertically. They lift away from their perch by sheer muscle power. We know that their wing muscles account for a quarter of their total body weight, but still that doesn't make it seem possible. Their hearts are enormous for their size because the oxygen they have to distribute while hovering is nothing short of extraordinary. The fact that they may beat their wings (including rotation and up- and-down-stroke) as many as 28.5 times every second was established a long time ago. Still, the only thing that makes the performance seem even remotely possible is that they actually do it! Hummingbirds fly.

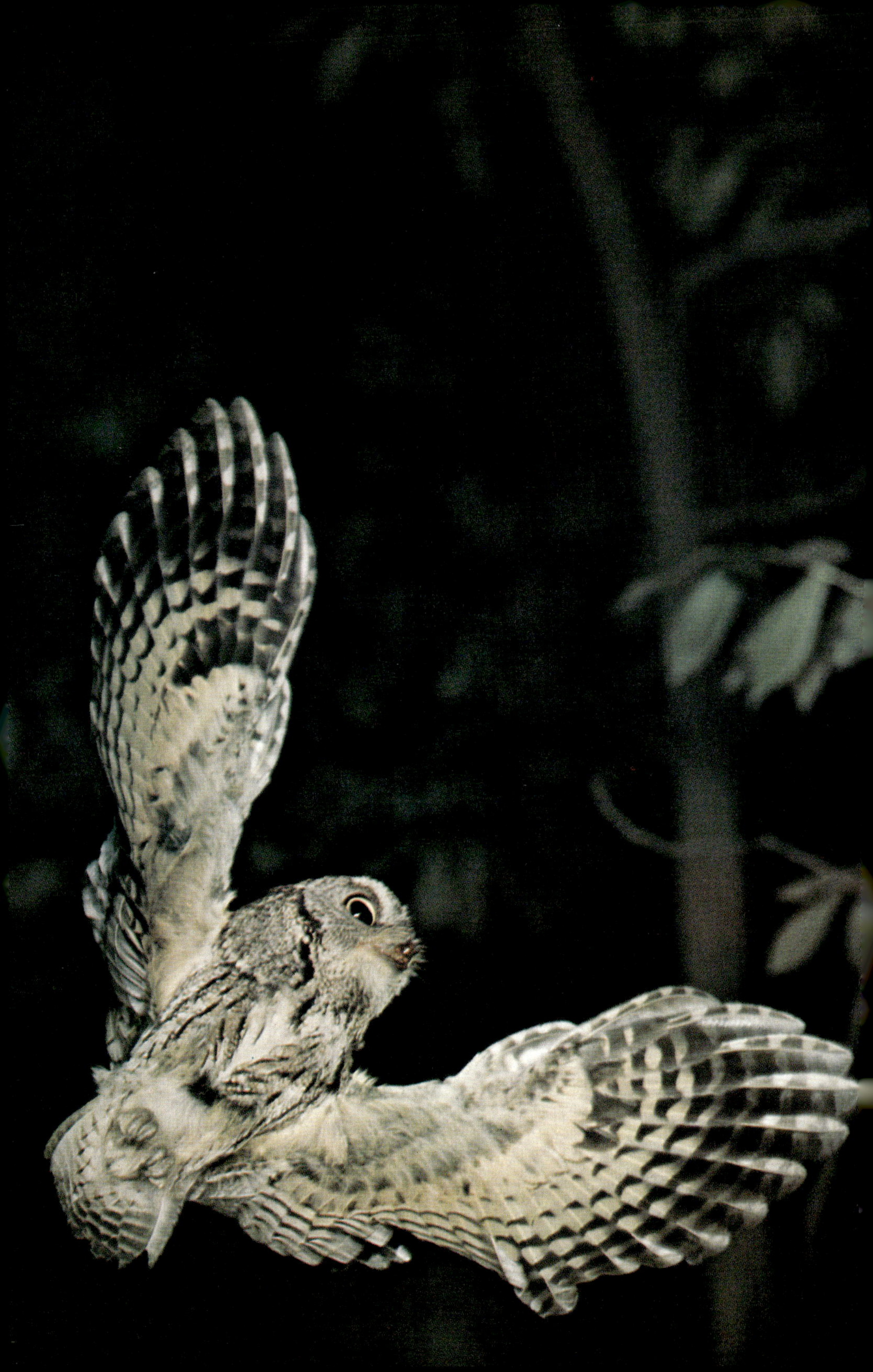

Owl in Flight

There is an undeniable fascination for us in the movements of an owl. First, it means death. Those of us who know birds know that. Owls don't fly for fun the way some birds do. An owl in flight is a bird on an errand and that errand is inevitably death—delivered personally. But that kind of death is justified because it is also life. No owl would be foolish enough to use up the energy it got from killing one animal by killing another it didn't need for food. Owls are smarter than we are in that one way at least. Perhaps that is the fascination—their intelligence.

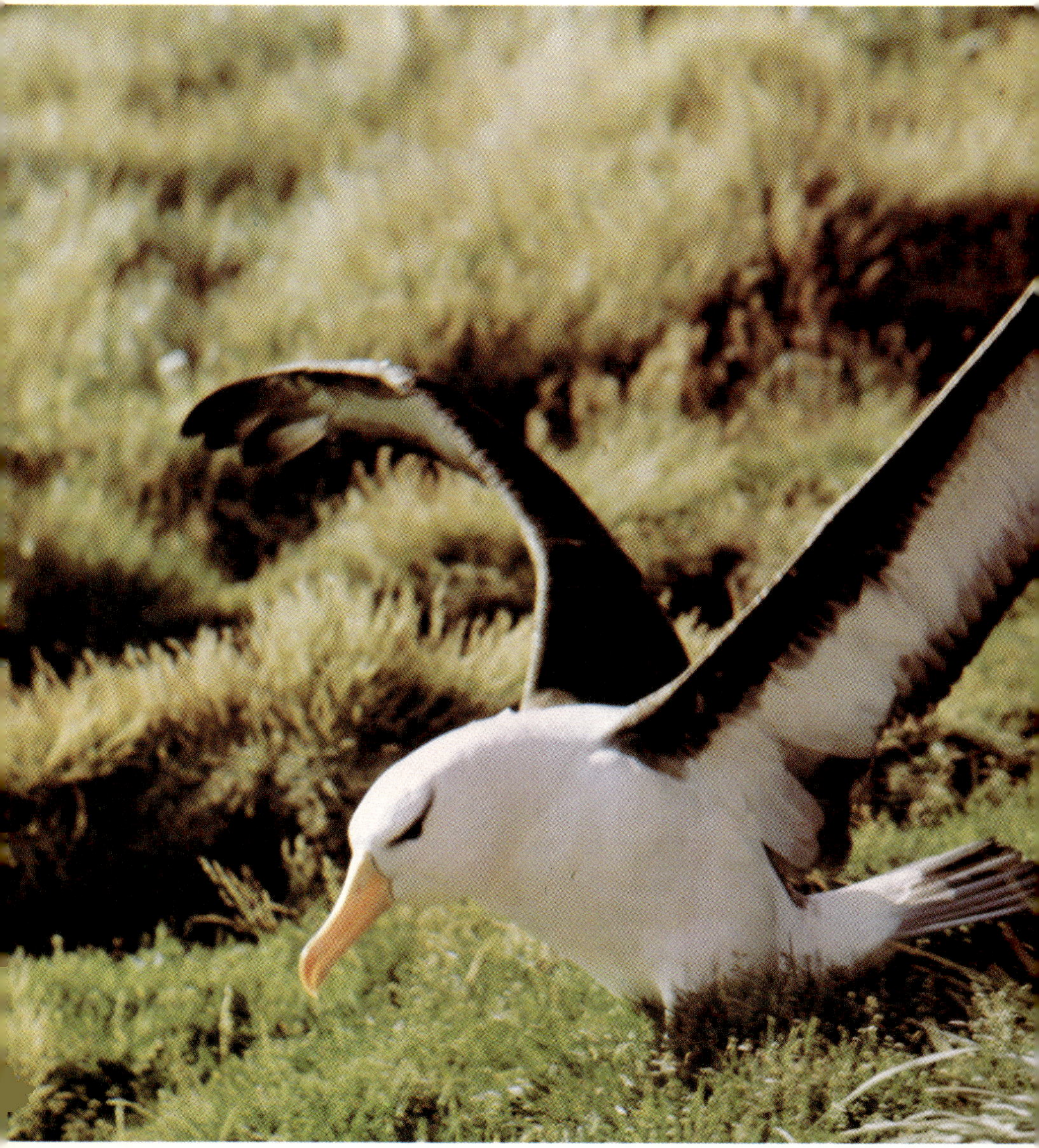

Black Albatross

I have stood for hours at the rail watching albatross tease the waves with the tips of their wings. To say there is no magic in their power is to admit you have never seen one. No corrugated sea is too intricate. The birds trace it, repeat its details inches away from the surface and then, without apparent effort, lift away to the sky, stand quite still, and then drop to taunt the waves again. The sea grabs at them eternally, and misses always. The enthralling sight of an albatross is the only cure I have ever found for the chronic seasickness with which I was blessed.

Macaw

There is, of course, a scientific explanation for why birds look as they do, why the bizarre colors and the strange shapes. It all can be explained in terms of genetics and species survival, of males attracting females, and camouflage. But, somehow, when you look at something like a macaw or one of the other "neon" birds, genetics isn't enough. It doesn't allow for the designer's sense of humor.

Galapagos Flamingos

At Lake Nakuru and at Hannington in Kenya, and later in the Ngorongoro Crater in Tanzania we saw assemblages of not thousands, but millions of flamingos. No matter where you have been, no matter what you have seen, you are not prepared for this. It is an experience from another world. Suddenly the flamingos start to fly. First a few, then a few hundred, then thousands, and then the sky is painted pink. The spectator finds himself standing with his jaw agape, staring wide-eyed at the sight and sound of a million flying birds laying the pink gauze of their flight across a land that seems black in contrast.

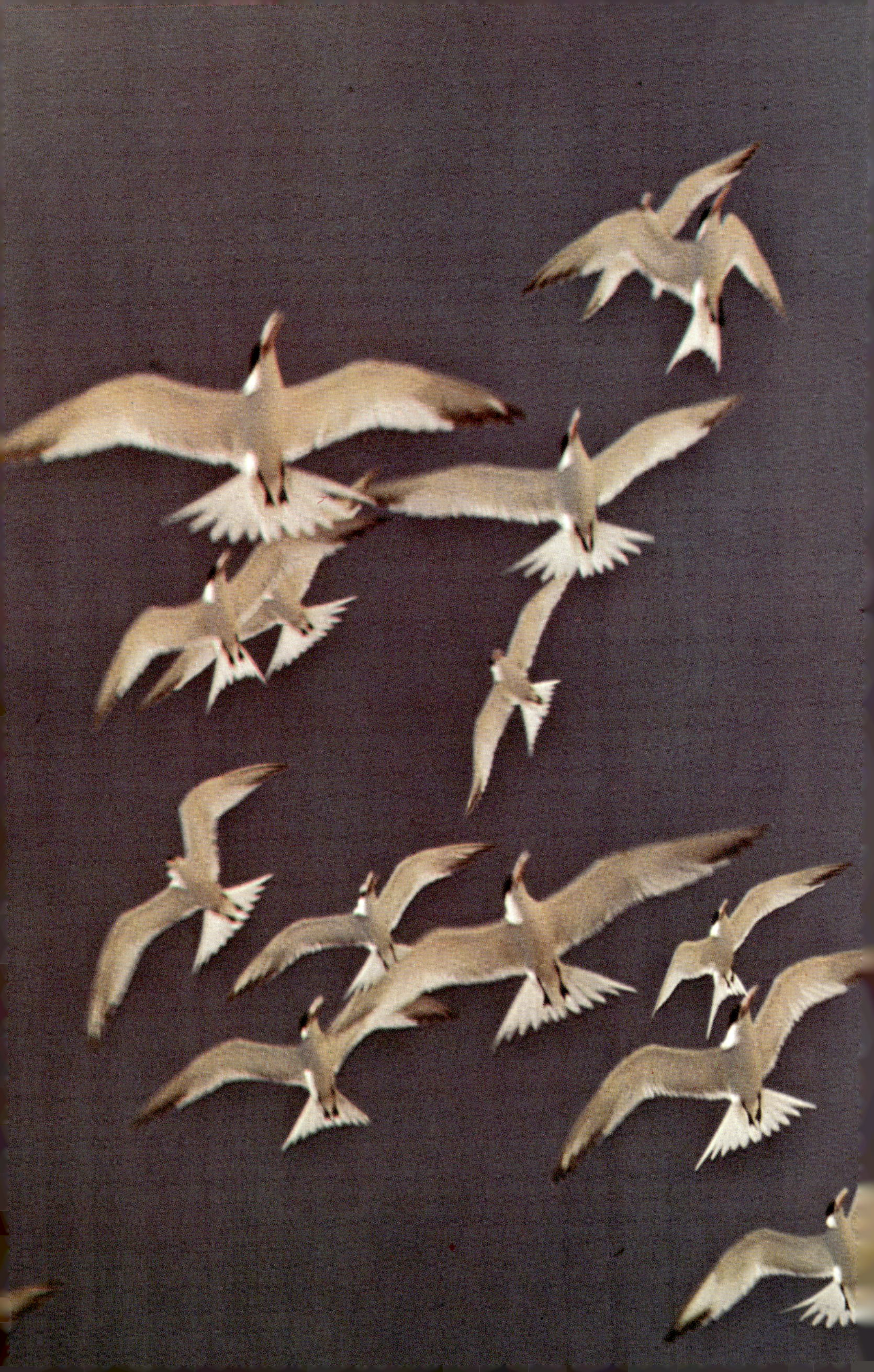

Flock of Terns
"When fishing the tern flies slowly, looking down into the water, occasionally glancing forward to check that the course is clear. On sighting a fish the tern dives with partly closed, angled wings, plunging like a white arrow into the sea. Sometimes, if in doubt, it will not dive immediately, but will hover over the spot where it suspects there is a meal, having made up its mind it will then either dive, or move slowly on in further search."

>From *Flying Birds* by
David and Katie Urry
(Harper & Row, New York, 1969)

Stork

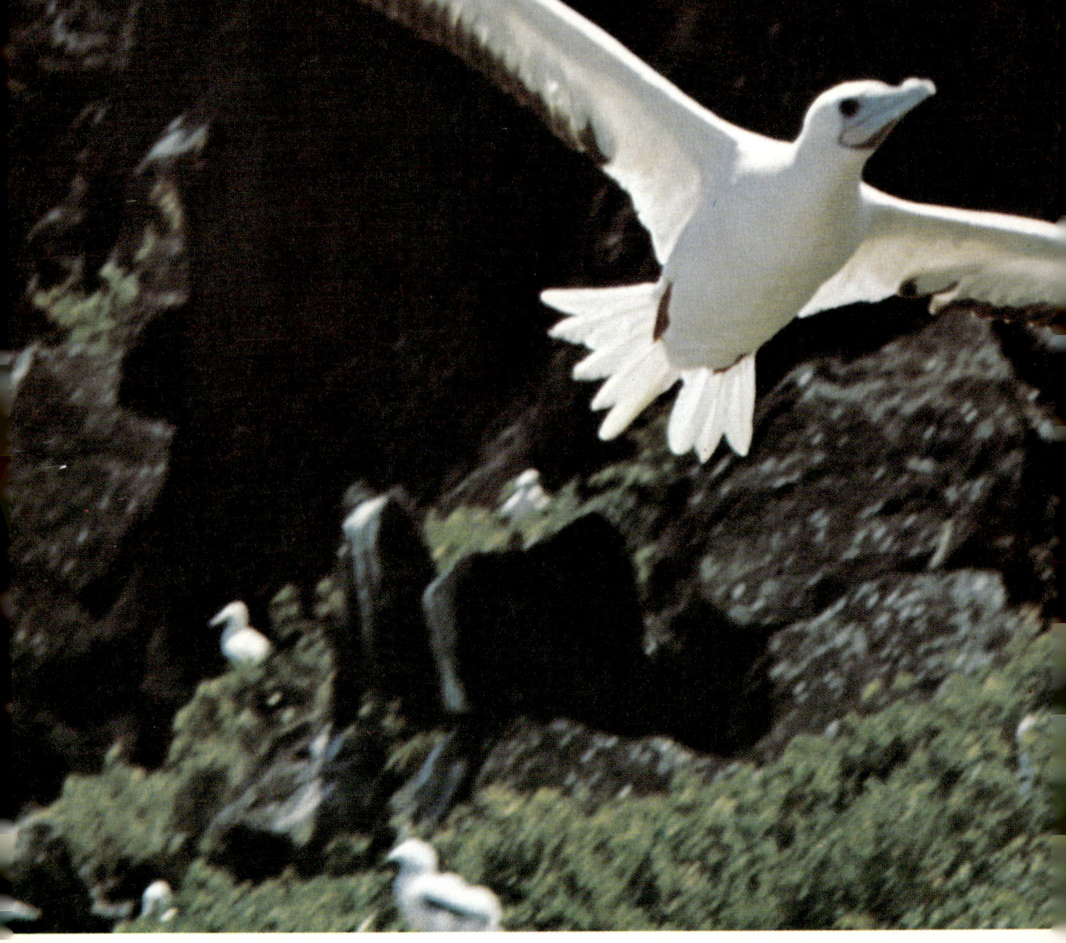

Red-Footed Booby

The power of flight has given birds a strange influence over the rest of the world. Birds, because of the distances they can cover, are *vectors*. Along with wind and water they account for plant distribution from mainland to mainland, from island to island. To some degree they are also responsible for the distribution of arthropods. In their digestive tracts and on their feathers birds carry life around the world. An undigested seed in the stomach of a bird is potentially the plant invasion of a distant land and the reshaping of the landscape.

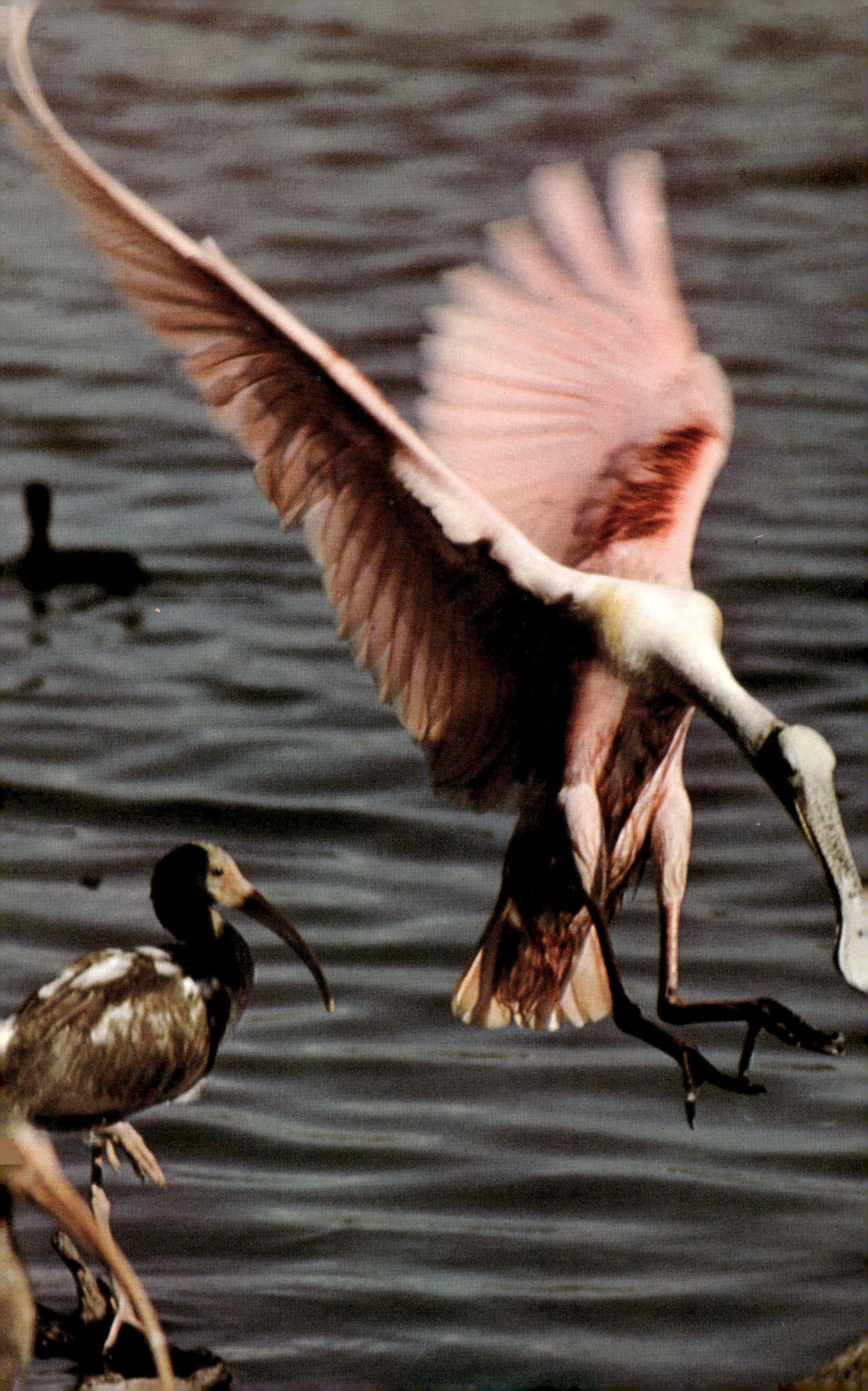

Roseate Spoonbill with a White Ibis

We seldom bother to think through the reasons nature had for introducing flight to the animal kingdom. It isn't only a means of getting from here to there but also a means of vertical distribution. Discounting marine and aquatic creatures, because they have both vertical and horizontal capabilities, all vertebrate animal life has horizontal distribution and that can make for crowding. Flight opens up the trees as well as the sky, and ledges, cliffs, and mountains. Flight also enables birds to migrate often over vast distances.

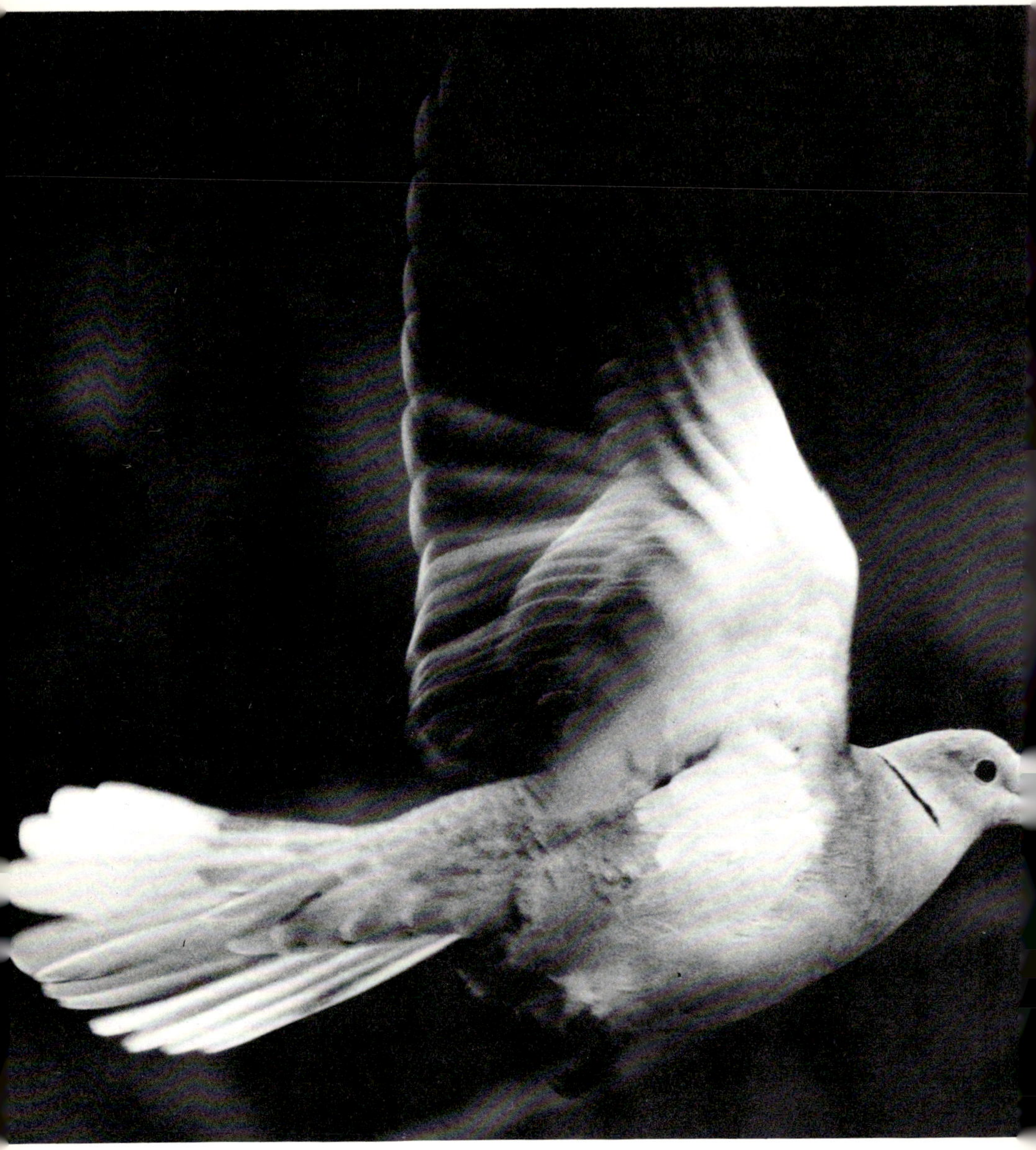

Collared Dove
We aren't jealous of an eagle's power of flight because we are in awe of the animal. We can forgive an ibis or an egret for taking wing because they are far lovelier creatures than we will ever even aspire to be. A soaring vulture—well, it is writing an important message in the sky. But this inferiority of ours really hurts when common birds such as pigeons and doves suddenly lift off and taunt us with the fact that they can do with ease what we can barely do with tons of fuel and machinery. That a dove and not a man can fly is a message of profound importance. I am sure we are being told that we have a place and we shouldn't forget it. Planes and rockets are fine, but a dove can really *fly!*

Wood Ibis

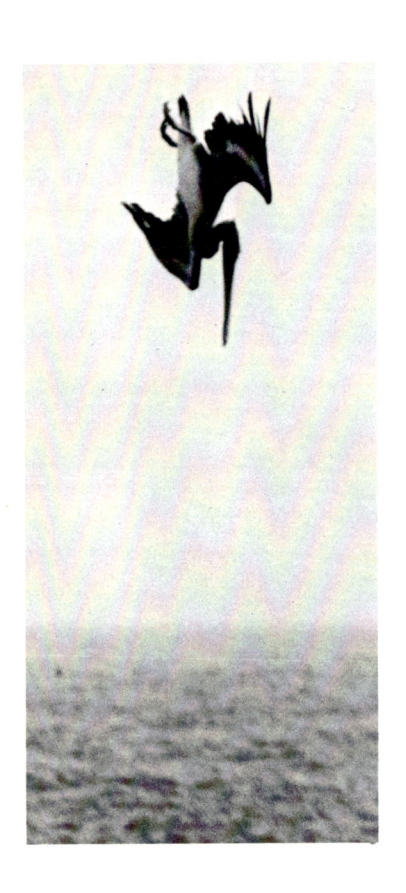

Brown Pelican Diving

"It is impossible to describe the beauty of their flight. . . . As one bird, the flock turns and dips and swoops toward the surface of the pond . . . then they swerve again, and the sunlight is reflected at exactly the same instant from every iridescent wing. As they veer sharply in front of us, the full spread of every individual bird's back and wings is turned toward us; then, almost between winks, the Spirit of the flock has brought the profile of each ibis in sharp silhouette against the sky—and half a hundred birds which seem like one with nine and forty shadows."

> by C. William Beebe
> From *Two Bird-Lovers in Mexico* (1905)

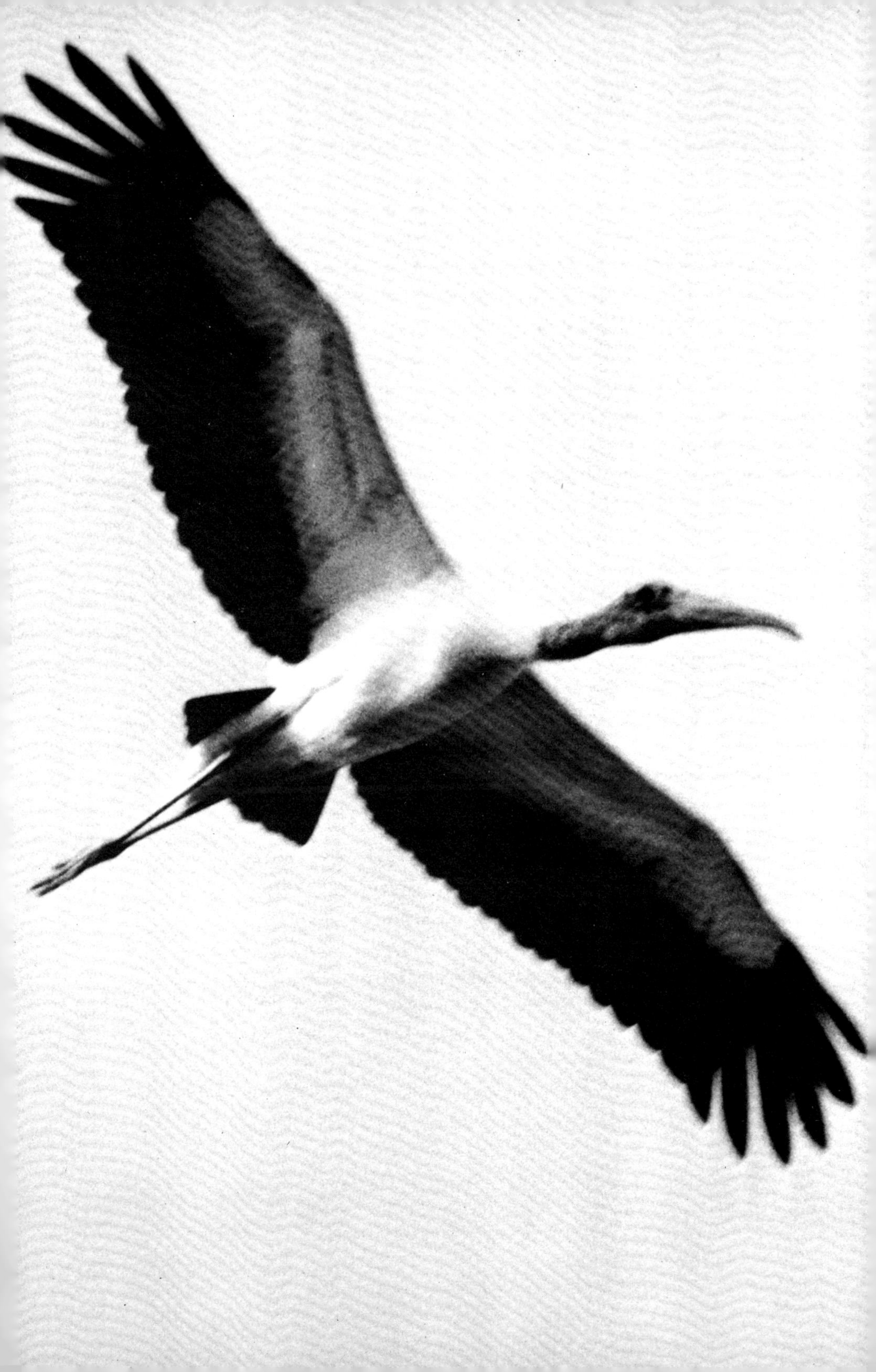

California Condor

"It is, perhaps, the cleanliness of the altitudes that would be most foreign to land-bound animals, the lack of clutter, of extraneous sound, and of debris, the utter freshness of it that would seem the strangest. Men feel the exhilaration of this new, clean world when they climb a mountain or soar silently in a glider. Perhaps the condor does too, for the young bird flew aimlessly, freely, going nowhere in particular, staying higher than was practical for him if he sought a morning meal. Perhaps the condor, too, can bathe his condor soul in the clean, fresh air of the altitudes. This community of spirit, if it exists, does not depend on manlike qualities in the bird, but on the bird in us."

>From *Source of the Thunder: The Biography of a California Condor*
Roger Caras (Little, Brown, Boston, 1970)

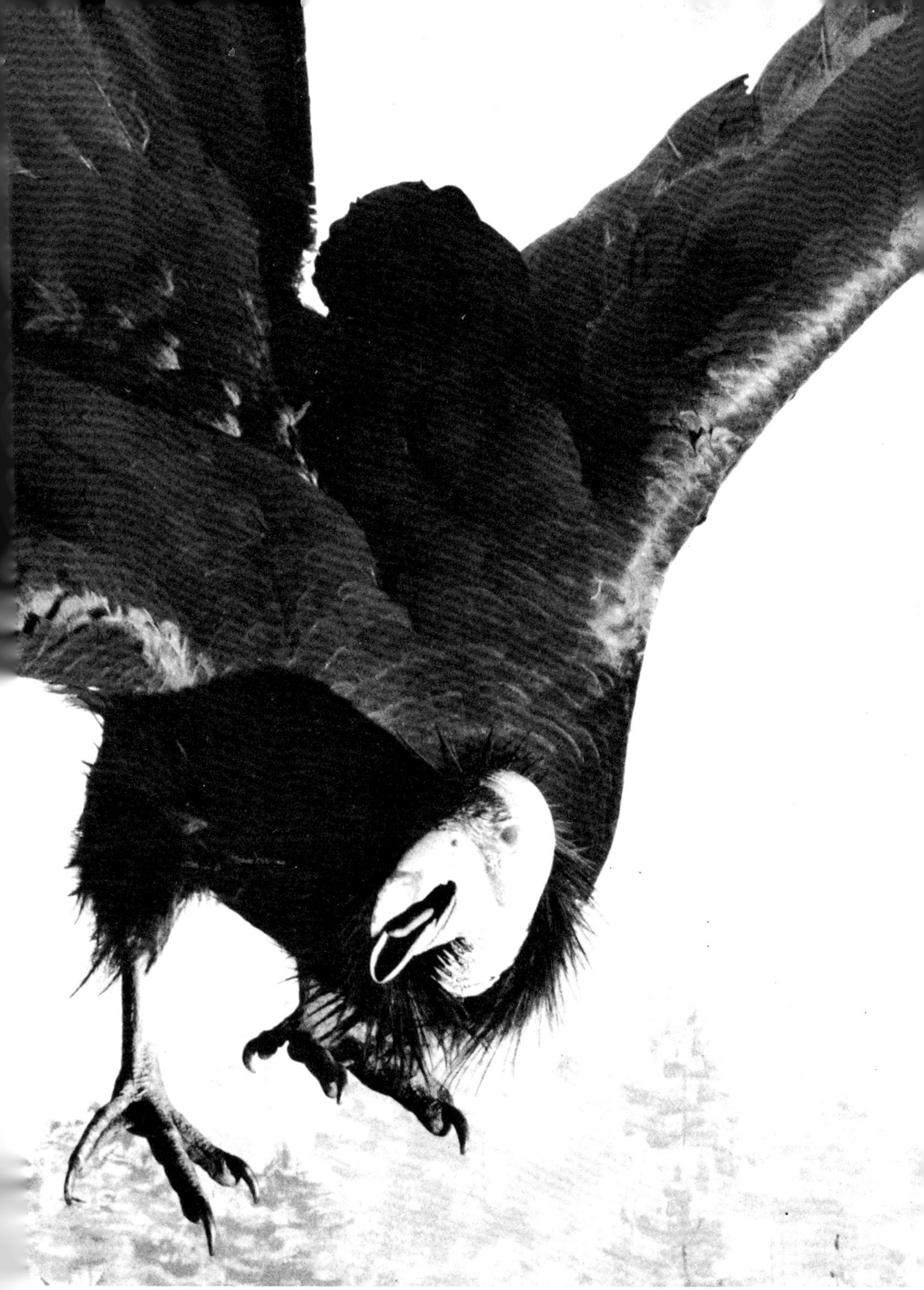

Reddish Egret

Rifleman

Nature certainly has not completed the design job. The enormous variety in bird form and flight style must be stages in the development of some absolutely ultimate perfection we can't even begin to envision. From the small and the blunt to the sinuous grace of an egret, a stork or a heron, some birds always manage to look frantically busy. Others, even when flying, somehow manage to look like Chinese art. Or is it that Chinese art once managed to look like the flight of wading birds? One way or the other the flight of birds reaches right down into the core of our aesthetic sense. A bird in flight is a new composition, a new look of art every second. A bird in the sky is a fluid painting and a sculpture of mist and sound.

Rufous-tailed Jacamar

Lappet-faced Vulture

"Birds have special devices built into their wings to increase lift and prevent stalling at low speeds. Equipment that functions on the same principles is used by airplanes in taking off and landing. When a wing reaches the stalling point, the stall can be controlled and the lift restored, without changing the angle of attack, by increasing the speed of air flow over the wing. This is accomplished by forcing the air to flow through a slot.... Birds use at least four: (1) slots at wing tips; (2) slots in front of the wing, made by the alula or by the first primary; (3) slots behind the wing, between the wing and the tail; (4) slots through the wing, made by lifting the covert feathers."

>From *The Flight of Birds* by John H. Storer (Cranbrook Institute of Science, Bloomfield Hills, Michigan, 1948)

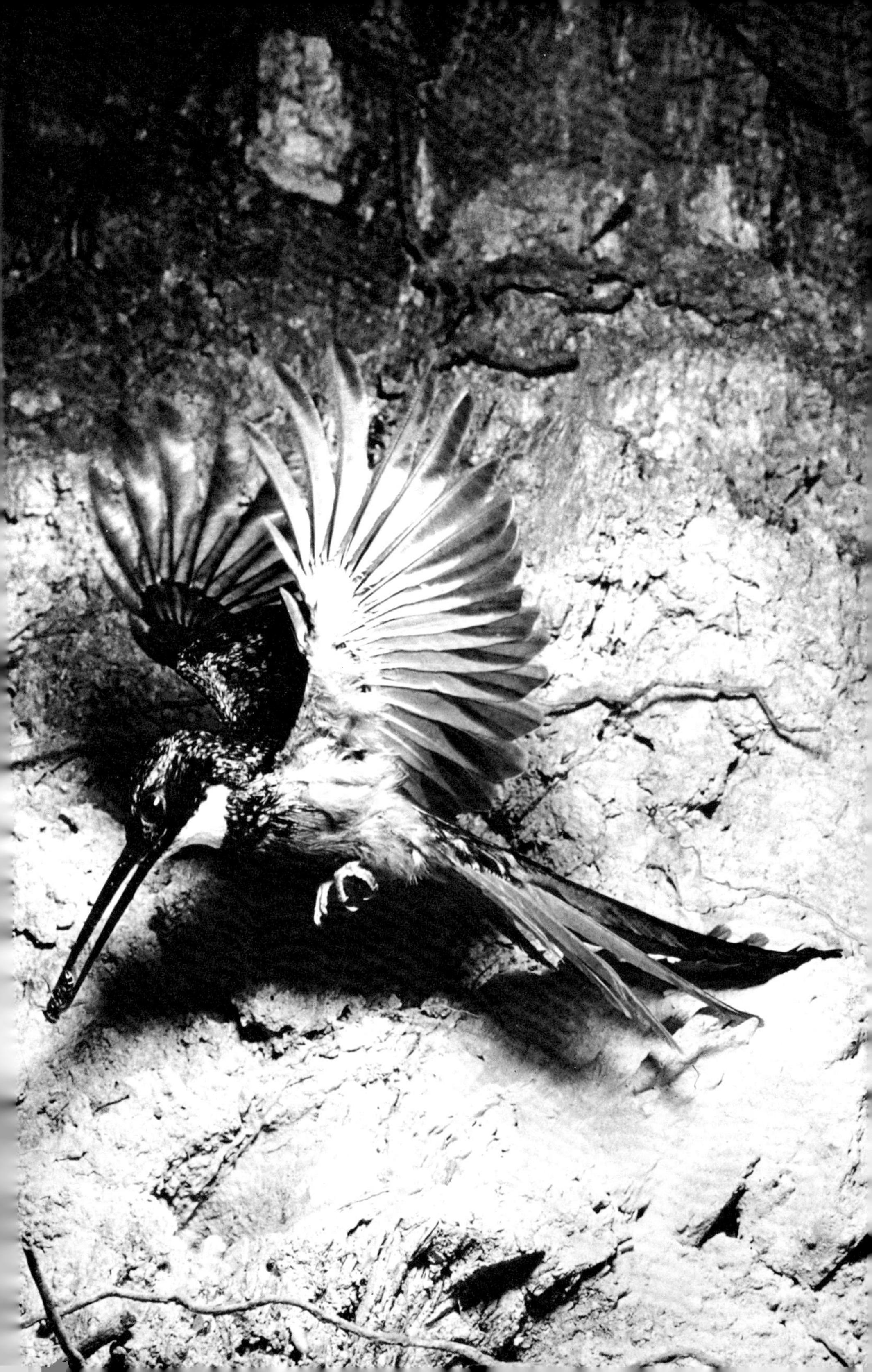

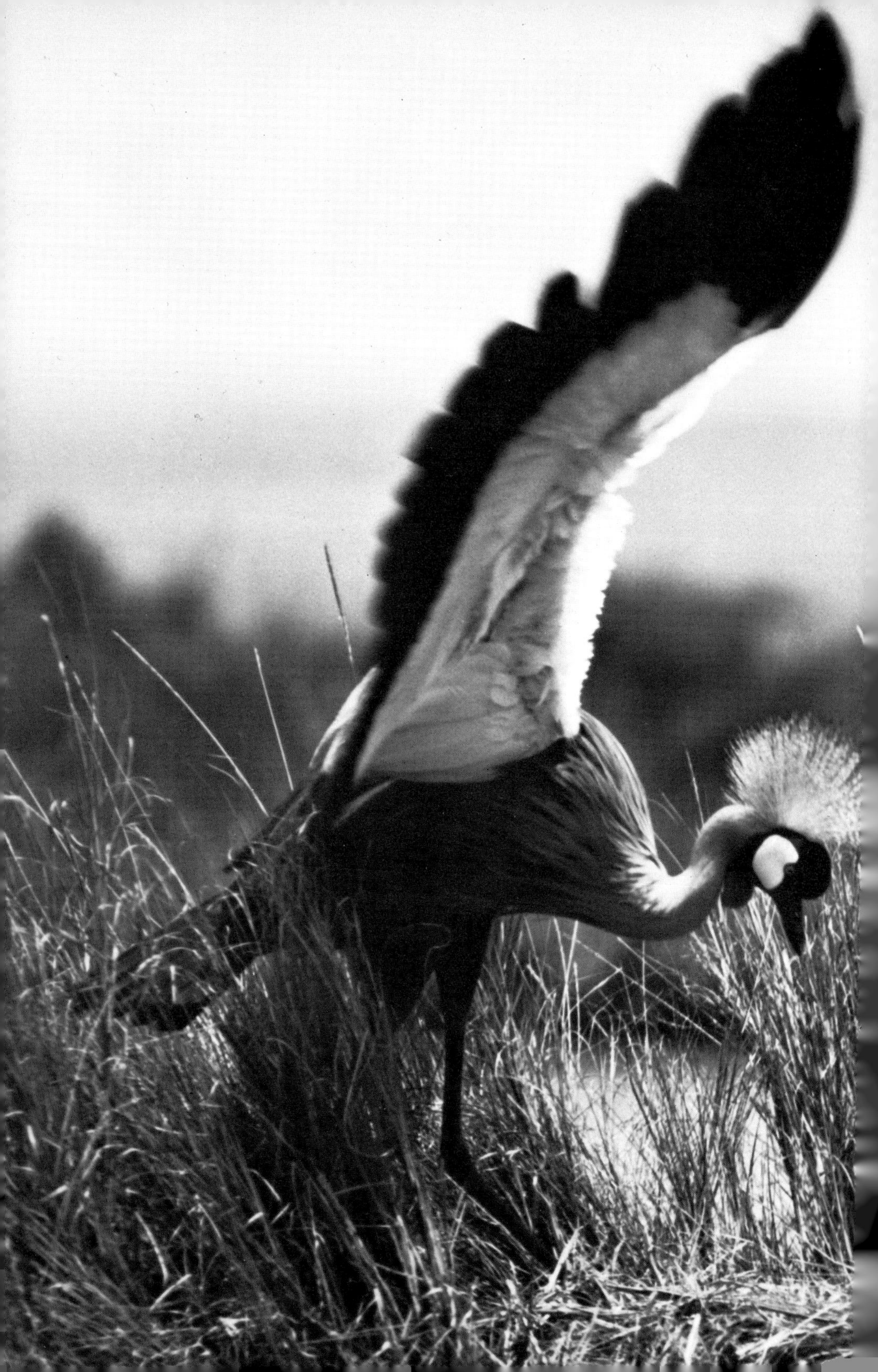

Crowned Crane

Why Nature felt compelled to give anything as magnificent as a crowned crane the power of flight along with all of her other gifts must remain a mystery. One would think that any creature with that much "going for it" would be considered rich enough without empowering it to lift above the earth and look down on all the rest of us. Somehow it just doesn't seem fair. Another case of the rich getting richer, I guess.

Brown Pelican

"Can you imagine any better example of divine creative accomplishment than the consummate flying machine that is a bird? The skeleton, very flexible and strong, is also largely pneumatic—especially in the bigger birds. The beak, skull, feet, and all the other bones of a 25-pound pelican have been found to weigh but 23 ounces.

    From *Song of the Sky* by
    Guy Murchie, Jr. (Houghton Mifflin, Boston, 1954)

Downy Woodpecker

Birds preceded man by millions of years. They reached their perfection before we found the threads of our beginnings. There never was a time and never a place that man, and pre-man for that matter, could not look up and see the flying animal. All through man's art, all through his literature this fact has been reflected. We have been profoundly affected from our earliest moments of awareness by birds and by flight.

We will never know just how much birds have had to do with our own intelligence, our own drive to know and understand. It is inconceivable that two million years of wondering about birds has not had an effect. Some slight hint of the depth of that effect can be found in the fact that after two million years and more we are still wondering. Perhaps early man was lifted out of the shroud of ignorance on the wings of birds. Perhaps birds have been our guides in more ways than we can ever hope to understand.

PHOTO CREDITS

PAGE	SUBJECT	PHOTOGRAPHER
5	Long-tailed Jaeger	Jerry L. Hout
6	Mallard Ducks	Toni Angermayer
8	Secretary Birds	Christina Loke
10	Mallards on the wing	Russ Kinne
11	Vulture	John G. Ross
12	Barn Owl	Karl H. Maslowski
14	Stork in flight	D. W. Friedmann
15	Gulls	Fritz Henle
16	Bald Eagle	Tony Kelly
17	Eagle in flight	Karl W. Kenyon
19	Pigeon Hawk	Karl H. Maslowski
20	Tufted Titmouse	Russ Kinne
21	Galapagos Hawk	Miguel Castro
22	Trumpeter Swan	Joseph Van Wormer
23	Trumpeter Swan	Russ Kinne
24	Canada Geese	Russ Kinne
25	Mute Swans	Russ Kinne
26	Fairy Tern	George Laycock
28	Black Skimmers	Russ Kinne
29	Peregrine Falcon	Lewis W. Walker
30	Frigate Bird	George Laycock
33	Laughing Gulls	Russ Kinne
34	Osprey or Fish Hawk	Russ Kinne
35	Steller's Sea Eagle	Russ Kinne
36	American Egret	J. H. Carmichael
37	Hummingbird	E. Moench
38	Owl in flight	Ron Austing
40	Black Albatross	Roger Tory Peterson
42	Macaw	Robert M. Eiselen
43	Galapagos Flamingos	Roger Tory Peterson
44	Flock of Terns	Roger Tory Peterson
45	Stork	Susan McCartney
46	Red-Footed Booby	Russ Kinne
48	Roseate Spoonbill with a White Ibis	James P. Jackson
50	Collared Dove	Russ Kinne
52	Brown Pelican diving	Al Lowry
53	Wood Ibis	James Hancock
54	California Condor	Tom McHugh (Field Museum)
56	Reddish Egret	Lewis W. Walker
57	Rifleman	Charles R. Meyer
58	Lappet-faced Vulture	Des Bartlett
59	Rufous-tailed Jacamar	Paul Schwartz
60	Crowned Crane	Des Bartlett
61	Brown Pelican	Al Lowry
62	Downy Woodpecker	Russ Kinne